IIW Collection

Series editor

IIW International Institute of Welding,
ZI Paris Nord II, Villepinte, France

About the Series

The IIW Collection of Books is authored by experts from the 59 countries participating in the work of the 23 Technical Working Units of the International Institute of Welding, recognized as the largest worldwide network for welding and allied joining technologies.

The IIW's Mission is to operate as the global body for the science and application of joining technology, providing a forum for networking and knowledge exchange among scientists, researchers and industry.

Published books, Best Practices, Recommendations or Guidelines are the outcome of collaborative work and technical discussions-they are truly validated by the IIW groups of experts in joining, cutting and surface treatment of metallic and non-metallic materials by such processes as welding, brazing, soldering, thermal cutting, thermal spraying, adhesive bonding and microjoining. IIW work also embraces allied fields including quality assurance, non-destructive testing, standardization, inspection, health and safety, education, training, qualification, design and fabrication.

More information about this series at http://www.springer.com/series/13906

Bastien Chapuis · Pierre Calmon
Frédéric Jenson

Best Practices for the Use of Simulation in POD Curves Estimation

Application to UT Weld Inspection

INTERNATIONAL INSTITUTE OF WELDING
A world of joining experience

Bastien Chapuis
CEA LIST
Gif-sur-Yvette
France

Pierre Calmon
CEA LIST
Gif-sur-Yvette
France

Frédéric Jenson
Safran Tech
Magny-Lès-Hameaux
France

and

CEA LIST
Gif-sur-Yvette
France

ISSN 2365-435X   ISSN 2365-4368   (electronic)
IIW Collection
ISBN 978-3-319-87365-7   ISBN 978-3-319-62659-8   (eBook)
https://doi.org/10.1007/978-3-319-62659-8

© Springer International Publishing AG 2018
Softcover re-print of the Hardcover 1st edition 2018
This work is subject to copyright. All rights are reserved by the Publisher, whether the whole or part of the material is concerned, specifically the rights of translation, reprinting, reuse of illustrations, recitation, broadcasting, reproduction on microfilms or in any other physical way, and transmission or information storage and retrieval, electronic adaptation, computer software, or by similar or dissimilar methodology now known or hereafter developed.
The use of general descriptive names, registered names, trademarks, service marks, etc. in this publication does not imply, even in the absence of a specific statement, that such names are exempt from the relevant protective laws and regulations and therefore free for general use.
The publisher, the authors and the editors are safe to assume that the advice and information in this book are believed to be true and accurate at the date of publication. Neither the publisher nor the authors or the editors give a warranty, express or implied, with respect to the material contained herein or for any errors or omissions that may have been made. The publisher remains neutral with regard to jurisdictional claims in published maps and institutional affiliations.

Printed on acid-free paper

This Springer imprint is published by Springer Nature
The registered company is Springer International Publishing AG
The registered company address is: Gewerbestrasse 11, 6330 Cham, Switzerland

Acknowledgements

This document has been significantly improved after reviewing by the following experts:

- John Aldrin, USA
- Charles Annis, USA
- Nicolas Dominguez, France
- David Forsyth, USA
- Daniel Kanzler, Germany
- Tony Walker, UK

The authors express their gratitude to them.

Contents

Introduction

Context

The quantification of the performances of a Non-Destructive Technique (NDT) is essential to guarantee the safety of critical industrial components. Two main indicators are commonly used:

- Reliability: the ability of the technique to detect defects under realistic conditions of application.
- Accuracy: the effectiveness of the technique to size the defect.

The estimation of Probability of Detection (POD) curves consists in a systematic methodology to evaluate the reliability of one inspection technique when applied to a defined class of parts. Meanwhile, the accuracy of the technique is evaluated establishing defect sizing curves (i.e. curves comparing size determined by the NDT to real size of defects obtained through destructive measurements), in particular to guarantee that the system does not undersize the defects.

It is the variability of conditions and of the parameters influencing the result of one inspection which motivates the use of probabilistic approach, and the POD study aims at answering this question: What is the probability that, at the end of one inspection performed by one operator applying the defined technique on one part belonging to the defined class and containing a defect of size a, this defect is detected?

Determination of POD curves via a purely experimental approach requires large-scale experiments performed on representative testblocks containing representative defects. The experiments, generally performed in lab, are designed in a way such that they account for the effects of the variability of influential parameters and of degraded conditions compared to nominal ones in real field. Practical feasibility, time and cost of such trials are today a major issue.

General documents exist to provide guidance on how to determine the POD curves. The MIL-HDBK-1823A [3], derived from several years of efforts in the USA to deploy the statistical methodology in the aeronautic industry, is the reference handbook. Relying on the same methodology framework, the ENIQ released several reports to introduce the methodology in nuclear industry in Europe [8]. In the oil and gas industry, the qualification of NDT is generally performed according to the DNV methodology [1, 2].

The use of numerical simulation should help the industry to determine POD curves for their inspection techniques in a more systematic way, with reduced time and cost. The use of numerical simulation to determine POD curves has been a subject of research of different industries and laboratories in the past years. The subject has been widely known under the acronym MAPOD for Model Assisted POD. The MAPOD approach is very attractive but still today suffer from a lack of a recognized and systematic methodology to be more widely used in the industry and accepted by the regulation authorities. In the latest version of the MIL-HDBK-1823A (2009), an appendix dedicated to the MAPOD approach has been added. However, this appendix only provides a general framework but no practical recommendations on how to deploy MAPOD approach in a performance demonstration program.

Scope of the Document

The aim of this document is to provide best practice guidance and practical recommendations on the use of numerical simulation for POD curve estimation in the study of NDT reliability. The document focuses on UT weld inspection but many of the principles could be applied to a broader range of techniques and situations. The use of simulation to evaluate the sizing accuracy (i.e. the NDT accuracy) is not described in this document.

Content

In the first part, the principal documents that establish the recommended statistical framework adapted for POD curve estimation are listed and briefly described. The most important initiatives on MAPOD approach in the recent years are also presented.

In the second part, the advantages and limitations of the simulation in this context are detailed.

In the third part, the prerequisites for the use of simulation (validation of the software, expertise of the user) are described.

In the fourth and central part, the methodology and guidance are sequentially described and possible applications for using POD curves determined using simulation are presented.

Part I
Background on POD and MAPOD Approach

Chapter 1
Recommendation Documents on Statistical Methodology Dedicated to POD Analysis

The literature on probability of detection is very large and the intent of this document is not to give a full review on the subject. It will, therefore, only indicate documents that give essential recommendations on the statistical methodology to setup and analyze a POD curve estimation campaign and documents and recent initiatives presenting the use of simulation in this context.

The statistical framework introduced in the documents below has been originally proposed to analyse POD data obtained from experiments. However, it can be directly applied to process POD data obtained from the simulation of the inspection method, and is therefore the recommended methodology to follow. It is derived from the original work of Berens in the 80s and based on the assumption of a functional form for the POD curve and maximum likelihood parameter estimates.

- **Georgiou, G., Probability of Detection (PoD) curves—Derivation, applications and limitations, *Health and Safety Executive report 454*, 2006**

In this report, Georgiou gives interesting and clear information on POD curves and their practical implementations and limitations. He especially warns about the validity of conclusions that can be drawn from a single POD curve and the danger to apply the curve to a configuration different from the one for which it has been determined. It is why it is very important to carefully report the input data of the simulations to clearly define the range of validity of the result (Part 4, section "Step 5: Evaluation of the Reliability of a Simulated POD Curve").

- **Annis, C., General guidance on how to determine POD curves, Department of Defense, USA, *MIL-HDBK-1823A*, 2009**

The MIL-HDBK-1823A, written by C. Annis, gives a general methodology for evaluating the performance of an NDT method in terms of POD for aeronautics industry. It is the reference document and a de facto standard in many industries. The statistical methods described in the Appendix G can be applied for analyzing the simulation data of ultrasonic weld inspection in order to determine POD curves.

© Springer International Publishing AG 2018
B. Chapuis et al., *Best Practices for the Use of Simulation in POD Curves Estimation*, IIW Collection, https://doi.org/10.1007/978-3-319-62659-8_1

It has to be noted that an appendix has been added in the latest version of the document (2009) giving conceptual guidance for the model-assisted estimation of POD (Appendix H).

- **Gandossi, L. Annis, C., Probability of Detection Curves: Statistical Best-Practices,** *ENIQ report 41*, **2010**

ENIQ report 41, written by L. Gandossi and C. Annis, gives a very good and clear overview of the justification of the best statistical methodologies used to produce POD curves. The statistical bases are the same as in MIL-HDBK-1823A.

- **Berens, A., NDE reliability data analysis,** *Nondestructive Evaluation and Quality Control,* **1989,** *17*, **689–701**

The three documents above rely on the initial work of Berens in the 80s, which introduced a statistical framework which is still considered as valid and recommended for the analysis.

Chapter 2
Documentation and Recent Initiatives on the Use of Simulation in the Context of POD

- **Matzkanin, G. & Yolken, T., *A Technology Assessment of Probability of Detection (POD) for Nondestructive Evaluation (NDE)*, NTIAC, 2001**

In this report Matzkanin and Yolken present a review of the early efforts towards the use of simulation for determining POD curves for different NDE methods until 2001.

The efforts since that date have mostly been supported through two initiatives: the MAPOD working group in the USA and the European project PICASSO. We will briefly describe these two initiatives.

- **MAPOD working group**

MAPOD working group, active from 2004 to 2011, gathered several entities in the United-States in order to promote the increased understanding, development and implementation of model-assisted POD methodologies.

More information about MAPOD working group and documents that it produced can be found at http://www.cnde.iastate.edu/mapod/index.htm.

- **PICASSO project**

PICASSO was a research collaborative project registered in the Seventh Framework Program of the European Commission from 2009 to 2012. It gathered several industrial partners of the aeronautics industry and research centers. The aim of the PICASSO project was to build a new and original concept of "POD curves based on NDT simulation in addition to existing experimental data base".

More information about PICASSO project can be found at http://www.picasso-ndt.eu/.

© Springer International Publishing AG 2018
B. Chapuis et al., *Best Practices for the Use of Simulation in POD Curves Estimation*, IIW Collection, https://doi.org/10.1007/978-3-319-62659-8_2

Part II
Advantages and Limitations
of the Simulation

Chapter 3
Advantages of Simulation

1. **Qualification time and budget saving**

   POD trials require large-scale experiments performed on representative components containing representative defects. Practical feasibility and cost of such trials is today a major issue. Using numerical simulation, the overall cost required to assess a POD curve can be significantly reduced. Depending on the objectives of such study (see Part 4, Chapter "Best Practice Principles and Guidance for the Exploitation of Simulated POD Data in Relation with Experimental Data"), the number of experimental testblocks (expensive to manufacture) necessary to obtain the information can be much fewer. In particular, the transfer function approach (Part 4, Section "Quantification of the Influence of a Change in the Inspection Procedure") allows, at a very little cost, to determine the POD curve for an inspection procedure derived from an initial one for which the POD curve has already been determined without performing a new set of experiments.

   Moreover, although the estimation of a POD curve by simulation generally requires a high number of simulations, the result is obtained much faster than when it is necessary to setup an experimental campaign.

   The simulation therefore makes the POD methodology more accessible and versatile than using a full experimental study. Consequently, this simulation based approach should help the industry to determine POD curves for their inspection techniques in a more systematic way.

2. **Possibility to generate large amount of data**

   Experimental campaign necessary to determine a POD curve are generally very expensive. Budget resources often limit the number of manufactured testblocks and the number of influent parameters tested. Using simulation, the number of influential parameters that can be introduced in the analysis can be more important which increases the robustness of the POD estimation.

© Springer International Publishing AG 2018
B. Chapuis et al., *Best Practices for the Use of Simulation in POD Curves Estimation*, IIW Collection, https://doi.org/10.1007/978-3-319-62659-8_3

The finite and somehow reduced number of experimental results is commonly considered as the most important factor influencing the accuracy of the POD estimation. Confidence bands classically account for this effect. In simulation studies, in general the statistical sampling is not a critical problem since it is possible to multiply the number of computations at practically no cost. Nevertheless, as discussed in Part 4 Section, "Step 5: Evaluation of the Reliability of a Simulated POD Curve" it is important to be aware that the statistical sampling is not the only possible source of inaccuracy on the POD estimation.

3. **Possibility to explore configurations not accessible experimentally**

A difficulty in experimental POD estimation campaigns is the manufacturing of realistic defects, for the whole range of defect sizes. Indeed, a correct estimation of the POD curve requires specimens with defects in the transition zone of the curve, but also tiny defects, that are not or barely detectable, and difficult to manufacture. The problem is even more crucial when it is required to implement these defects at a precise location in complex geometries. Simulation easily allows to explore these complex but realistic configurations and therefore to be confident on the reliability of the POD curve that is obtained. There are generally no limits in terms of configurations that can be considered in simulation campaigns, as long as the models are validated (Part 3, Chapter "Validation of the Software Tools").

4. **Interpretation of results**

Simulation can provide tools (beam computation, movies of ultrasonic wave interaction with defect...) offering very rich information which helps to understand the physical processes underlying the inspection process. This can allow to optimize the inspection setup or the procedure.

In the context of POD curve estimation campaigns this is of practical interest since simulation can allow to quantify the influence of the different aleatory parameters, and therefore to identify the parameter on which to focus on field in order to guarantee the POD (Part 4, Section "Modify the Procedure to Bring Back a POD Curve to an Acceptable Value").

Chapter 4
Difficulties, Limitations

Human Factors

A reluctance to use simulation for POD curve estimation is often due to the, alleged, inability of simulation to correctly reproduce the influence of human factor on the POD curve. The aim of this paragraph is to precise what is meant by human factors and what can be expected from simulation studies.

During several workshops dedicated to reliability in NDE ("European-American Workshops on Reliability of NDE" [24]), a modular model was proposed to address the various factors influencing the reliability (R) of a NDE system. This model recognizes that an NDE system has a certain intrinsic capability (IC) which can be reduced to application related parameters (AP) or human and organizational factors (OHF). This can be expressed using the following general expression:

$$R = f(IC, AP, OHF).$$

The IC is related to the physical principle behind the defect indication and its technical realization. AP includes the influential factors resulting from the realistic circumstances under which an inspection is performed.

Generally, organizational and human factors refer to all the parameters that need to be controlled to obtain reliable human performance [22]:

- Human (qualification, personality)
- Team (behavioral standards, norms, stress, leadership, communication)
- Technology (design)
- Organization (structure, procedure)
- Environment (regulator, manufacturer)

Health and Safety Executive (UK) website provides interesting classifications of the different sources of human factors in the industry [23].

© Springer International Publishing AG 2018

B. Chapuis et al., *Best Practices for the Use of Simulation in POD Curves Estimation*, IIW Collection, https://doi.org/10.1007/978-3-319-62659-8_4

In the context of NDE, AP gathers the part of the influence of the operator on the inspection that is directly related to the application of the procedure: manual positioning of the ultrasonic probe, interpretation of the signal by the operator, etc. The effects that result of unreliable human performance (due to the stress or to the fatigue of the operator, for example) are included in the term HF.

Usually experimental POD studies based on round robin trials, include trials performed by several operators. This allows capturing, but only partially, some aspects linked to HF. Only large scale experiments (such as PANI projects in UK [28]) have studied these effects.

Although some possibilities to include human factor influence in simulation POD studies are evoked in literature, this remains a topic of research. At the present stage of knowledge such inclusion cannot be expected by a simulation study.

The aim of this document is to address methods permitting the use of numerical simulation results in the process of POD estimation. Consequently, organizational and human related factors and the way to include those factors in the process of POD estimation will not be addressed in this document.

Lack of Knowledge on Influential Parameters, Epistemic and Aleatory Uncertainties

In the introduction we have recalled that the POD approach aims at taking into account the variability of the influential parameters. Let us recall that in the ENIQ terminology the parameters influencing the results of an inspection are called influential parameters or essential parameters [25]. This distinction refers to the degree of importance of this influence. Essential parameters are those which have to be taken into account in the qualification process.

Let us also precise the distinction between variability and uncertainty [26]: The uncertainty on an influential parameter reflects the lack of knowledge on the value it takes while its variability reflects the fact that this value may be different from one inspection to the other. The variability on influential parameters can be seen as one of the causes of the uncertainty on these parameters. Obviously there is always an uncertainty attached to every influential parameter even if it is controlled (in that case it is an experimental uncertainty).

At this stage it can be useful to introduce the concepts of "aleatory" uncertainty and "epistemic" uncertainty (see Ref. [27]). "Aleatory uncertainty is one which is assumed to be intrinsic randomness of the phenomenon" under consideration while "epistemic uncertainty is the one that is presumed as being caused by lack of knowledge". It has to be mentioned that obviously this distinction is subjective and depends on the context since more fundamentally, every uncertainty reflects a lack of knowledge.

In our NDT context the "aleatory uncertainty" corresponds to the variability of the influential parameters. It is this uncertainty, intrinsic to the NDT process under consideration, which we want to account for by a POD assumed to be an "intrinsic" property of this NDT process.

The "epistemic uncertainty" in our context is linked to our lack of knowledge on the inspection process (parameters and physical phenomena).

The "epistemic uncertainties" are not assumed to be accounted for by the POD but they may affect the accuracy of the estimated POD. It is the case if the epistemic uncertainty refers to a lack of knowledge on a quantity involved in the estimation of the POD.

Estimation of POD by simulation to be properly achieved requires more knowledge than estimation of POD from experiments. Therefore estimation of POD by simulation is more subject to "epistemic" uncertainty. This epistemic uncertainty specific to simulation mainly refers to:

- The lack of knowledge on influential parameters not submitted to variability which are inputs of the simulation code.
- The lack of knowledge on the variability of the aleatory parameters (influential parameters subject to variability).
- The uncertainty linked to the use of non-perfect models which can be seen as reflecting a lack of knowledge on the physical phenomena involved in the NDE process.

Identification of Aleatory Parameters

In the context of simulation it is necessary to identify all the aleatory parameters. Only the aleatory parameters that have been correctly identified (and for which there is a physical model linking the inspection response to the parameter value) will be taken into account by the POD. It may be difficult to identify all those parameters.

The identification should be performed with the help of an on-field specialist of the inspection method.

This identification is less critical in the case of experimental studies (round robin), since in that case it can be presumed that some "aleatory" parameters not identified primarily when defining the test matrix can contribute to the final result.

Knowledge of Statistical Distributions of Aleatory Parameters

The estimation of POD by simulation requires the knowledge of the statistical distributions describing the variability of the identified aleatory parameters.

For some parameters, these distributions can be difficult to obtain (it might require experimental measurements or knowledge of an on-field expert). A lack of knowledge on these distributions corresponds to an "epistemic" uncertainty which induces an uncertainty on the estimated POD. When the distributions are determined with a low confidence degree, it is recommended to perform a sensitivity study on the estimated POD as described in Part 4, Section "Quantification of the Influence of a Change in the Inspection Procedure".

It has to be noted, however, that in the case of experimental POD estimation, knowledge on the variability of influential parameters may also be required in order to ensure that the design of experiments (DOE) correctly sample the distributions of aleatory parameters. Therefore, in that case, the lack of knowledge on the distributions may also induce uncertainty on the POD. However, this uncertainty less critical than in the case of simulation studies and difficult to quantify is generally neglected.

Availability of Suitable and Validated Models

The models that simulate the inspection process must be validated in the full range of use as described in Part 3, Chapter "Validation of the Software Tools". The development and the validation of such models can be long and difficult in some situations.

Chapter 5
Possible Ways of Using Simulation in the Context of POD Curve Estimation

- NDT performances assessment at feasibility stage

  - Simulation allows to estimate the POD curve during the development of the inspection method. This provides good indications on the expected performances at an early stage of the project, for example during the ultrasonic probe optimization.

- Optimization of the design of experiments (Qualification/Validation)

  - The manufacturing of representative testblocks is an expensive step of an experimental campaign for POD curve estimation. Simulation can be used to select the trials (manufactured defects) in order to focus on the range of interest of the POD curve and therefore to reduce the number of specimens in ranges where POD = 0 or 1.

- Quantification of the effect of the variability of additional parameters

  - When a POD curve has been determined by an experimental campaign, the experiments cannot be reused if the variability of another influential parameter must be taken into account. In that case a full experimental campaign must be performed including the new parameter. Simulation can be used to avoid this new campaign by quantifying the impact of the new parameter.

- Identification of parameters for improvement of POD results

  - If the POD curve determined either by experiments or by simulation is not satisfactory (e.g. the $a_{90/95}$ value is too high), simulations can be used to determine which parameters of the procedure contribute significantly to the degradation of performance of the inspection method. Corrective actions to the procedure (like ensuring less uncertainty in probe positioning) can therefore be proposed and tested before renewing the experimental campaign.

© Springer International Publishing AG 2018
B. Chapuis et al., *Best Practices for the Use of Simulation in POD Curves Estimation*, IIW Collection, https://doi.org/10.1007/978-3-319-62659-8_5

- Complement experimental data by simulated ones to compute a full POD curve with better reliability

  - Simulation can be used as a complement of experimental data to compute a full POD curve. This can be done to increase reliability of the POD curve determined experimentally or to complement a limited number of experiments that was insufficient to compute the POD curve. The reliability is increased thanks to the better knowledge of the inspection process given by the complementary experimental and simulation approaches.

- Provide technical justifications when minor changes of the procedure occur

  - Simulation can be used as a technical justification when minor changes have been made to the inspection procedure avoiding a new experimental campaign.

- Design an inspection procedure with an objective in terms of POD

  - When the inspection method must be qualified it could be interesting to use simulation of the inspection procedure and optimization algorithms with an objective of POD performances.

- Worst case identification

  - Simulation can be used to identify the worst case of the inspection procedure, in order to concentrate the design of the inspection method on this case to guarantee a correct POD curve at the end of the experimental campaign.

- Training and evaluation of operators performance

  - Simulation can be used to train people involved in POD curve estimation campaign to understand the POD methodology. Simulation can also be used to generate signals and images that can be used to evaluate operators' performances.

Part III
Prerequisites for the Use of Simulation

Chapter 6
Validation of the Software Tools

The estimation of POD curves using simulation requires three major phases, namely:

(1) The generation of a set of input configurations assumed to be representative of the inspection conditions encountered in reality. This phase is called "design of (numerical) experiments" DOE even if in this case experiments are numerical ones.
(2) The simulation of the measurement process for this set of configurations.
(3) The statistical processing of the results obtained in the second phase.

The accuracy of the obtained POD curve depends on the validity of these three phases.

The question of the representativeness of the DOE (phase 1) is central, the POD being obtained either by simulation or by experimental studies. There is no simple way to a priori evaluate this representativeness nevertheless. However, recommendations concerning this question are given in Part 4, Section "Step 5: Evaluation of the Reliability of a Simulated POD Curve" of this document.

The phases 2 and 3 can be performed in the same simulation platform or through a coupling of two codes, one for the simulation itself (phase 2) and one for the statistical processing (phase 3). In any case, to guarantee a correct result, the two phases must be carefully and independently validated.

- **Validation of the simulation**
  One strong requirement when using simulation is to evaluate the accuracy of the predictions given by the code. It is particularly true in the case of the use of simulation for the calculation of POD curves. To produce consistent results, the physical models used in the simulations should be validated for the range of variation of the different aleatory parameters.
  The question of the validation of NDT models is discussed in the IIW Best recommended practice "Recommendations for the use and validation of non-destructive testing simulation" [7]. It is recommended to validate simulation

© Springer International Publishing AG 2018
B. Chapuis et al., *Best Practices for the Use of Simulation in POD Curves Estimation*, IIW Collection, https://doi.org/10.1007/978-3-319-62659-8_6

codes by comparing the predictions provided by the code with experiment. In [7] are listed the different possible sources of discrepancy between experiment and simulation. Among them are distinguished those which concern the model itself (and its implementation) and those which concern the representativeness of the parameters inputted in the code to account for the experimental conditions. In the context of ultrasonic testing, due to experimental uncertainties, a discrepancy between experiment and simulation less than 3 dB is considered as satisfying.

Nevertheless it is important to check if this measure of the model validity includes or not the uncertainty linked to the input of the code. If the experimental validation of the model has not been performed on mock-up identical to real components and using inspection devices identical to the ones used for the inspection under study, then the validity of the description of the parameters inputted in the code have also to be considered (see Part 4, Section "Step 5: Evaluation of the Reliability of a Simulated POD Curve").

- **Validation of the statistical algorithms**
  The statistical algorithms used to process the simulation results in order to determine the POD curve must also be used in their range of validity and their implementation validated. When using commercial package or an open source program, all reasonable precautions should be taken to be sure of the correctness of the software. When using in-house software it is recommended to validate the implementation of algorithms by comparing the output of the code to output produced by the same algorithm implemented in another software program already validated.

It has to be mentioned that the statistical algorithms used to process simulated data are the same as those that are used to process experimental data.

Chapter 7
Training/Expertise of the Operators

In order to ensure the correctness of results obtained thanks to simulation the operator responsible of the simulation campaign must have the following skills and knowledge:

- Use of the simulation software
- Knowledge on the inspection method that is simulated and its practical implementation on field
- Knowledge of the range of validity of simulation models and their application conditions
- Knowledge on the hypothesis underlying the statistical methodology of POD curve estimation.

© Springer International Publishing AG 2018   21
B. Chapuis et al., *Best Practices for the Use of Simulation in POD Curves Estimation*, IIW Collection, https://doi.org/10.1007/978-3-319-62659-8_7

Part IV
Methodology and Guidance

Chapter 8
Basics of Statistics for POD

POD Curves

This is not the aim of this document to present the full statistical framework for the POD curve estimation. As mentioned above, the methodology is the same as for analyzing data obtained experimentally and this methodology is described in detail in the MIL-HDBK-1823A handbook (Appendix G) [3] and the ENIQ report 41 [8]. However, some key concepts necessary to understand the next sections of the document are introduced here.

Let us consider the following relationship linking the response of an inspection system to the influential parameters of the inspection method:

$$y = f(a, X)$$

where:

- y is the response of the inspection system,
- a is the characteristic parameter of the defect (e.g. flaw dimension),
- X are the uncertain parameters.

The POD curve represents the relationship POD(a) that links the probability of detection to the characteristic parameter of the defect. The characteristic parameter is generally given by a mechanical resistance analysis of the structure under inspection and should be representative of the harmfulness of the defect.

Figure 8.1 shows (in blue) a typical POD curve. For low defect sizes the POD is 0, no defect of these sizes are detected. On the contrary, for large defect sizes the POD is 1, all the defects of these sizes are detected. The intermediate zone, generally for defect sizes of industrial interest, is a transition zone on which the POD curve increases from 0 to 1. This "S" shape of the POD curve is an assumption that is a good approximation of the experimental observations in the majority of the situations. The recommended methodology for determining the POD curve consists

© Springer International Publishing AG 2018
B. Chapuis et al., *Best Practices for the Use of Simulation in POD Curves Estimation*, IIW Collection, https://doi.org/10.1007/978-3-319-62659-8_8

Fig. 8.1 Typical POD curve
(in *blue*) and 95% confidence
limit (in *red*)

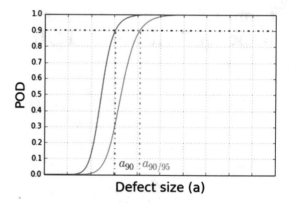

in using a model where the POD curve is assumed to have a functional form (of "S" shape) which depends on a small number of parameters (usually 2) [8]. These parameters are identified so that the functional best fits the inspection data, which is achieved using the maximum likelihood estimation method. This general methodology is called "parametric modelling" and has several advantages among the historical binomial approach (and the related "29 of 29 method"), whose use is now discouraged according to the ENIQ report 41. A particular care should, however, be taken to verify the hypothesis on the functional form, i.e. that the determined parameters give a model correctly that fits the data [31]. Other methodologies such as nonparametric models [32] might be used when this is not the case, but this is still a subject of research and beyond the scope of the document.

The POD curve is associated with a confidence interval. This confidence interval is introduced to account for the fact that the POD curve is an estimation based on a finite number of data. The meaning of a confidence interval at α% on POD(a) is the following:

Provided that the statistical model is correct then, the probability that the true value of POD(a) be in the confidence interval at α% is equal to $\alpha/100$. By true value, we mean the value of the "intrinsic" POD we attempt to determine and which account for the variability of influential parameters.

The important point to be aware of lies is in the condition «provided that the statistical model of data is correct». In particular, in the case of signal response analysis (cf. below), by correct we mean, not only that the Berens hypothesis are verified, but also that the statistical sampling (driven by the design of experiment) correctly capture all the variabilities of the inspection.

The second curve (in red) in Fig. 8.1 indicates therefore the lower bound of the confidence interval on the estimation of the POD curve. In practice α is generally chosen at 95%, the red curve is called the confidence limit at 95%. The defect size for which the POD curve reaches the values 0.9 is called a_{90}, which is therefore the minimal defect size detected 90% of the time. The defect size for which the confidence limit curve reaches the values 0.9 is called $a_{90/95}$. $a_{90/95}$ is now commonly used in several industries (aeronautics [3], oil and gas [1] for example) and is, de facto, the standard value to characterize the performances of the NDT system. Since $a_{90/95}$ is always higher than a_{90}, it is considered as a conservative value for a_{90}.

In accordance to the above remarks on the meaning of the confidence interval it is important to be aware that the confidence bound reflects only partly the uncertainty on the POD values. There are various epistemic sources of uncertainty in the process of estimation which are not linked to the statistical sampling. These epistemic uncertainties are of different nature for experimental and simulation POD as discussed in Part 2, Chapter "Difficulties, limitations".

Hit/Miss Analysis

In some cases the response of the inspection system is a binary response: the defect is detected or not, $y = 1$ or 0. This type of data is call "hit/miss". It can be, for example, the result of an inspection using TOFD system, for which an operator analyzes the image and from that concludes either he sees a defect or not.

An example of hit/miss data and the associated POD curve is presented in Fig. 8.2. The functional "logit" (also called "logistic" or "log-odds") is the most commonly chosen functional to fit the data; it has an "S" shape and can be parameterized by two parameters μ and σ, where μ indicates the flaw size which is detected with a probability 50% (POD = 0.5) and σ the steepness of the function (Fig. 8.3):

$$\mathrm{POD}(a) = \left[1 + \exp\left(-\left(\frac{g(a) - \mu}{\sigma}\right)\right)\right]^{-1}$$

where $g(a) = a$ or $g(a) = \log(a)$.

Fig. 8.2 Example of POD curve for hit/miss data (*blue points*)

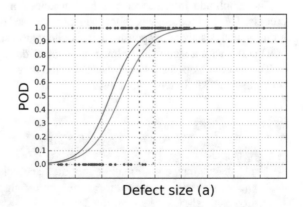

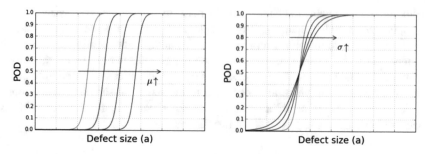

Fig. 8.3 Physical interpretation of the parameters μ and σ for the logit functional

The parameters μ and σ that best fit the data are determined by applying the maximum likelihood estimation methodology as described in ENIQ report 41, Sect. 3.4.3.

We also refer to the ENIQ report 41, Sect. 3.4 for a detailed explanation of this choice and other important considerations (logarithmic or linear regression, choice of the appropriate regression functional).

The methodology described in ENIQ report 41, Sect. 3.4.4 using the log-likelihood ratio should be used to determine the confidence band. Its approximation described by Cheng [9] is an acceptable approximation for large (>200) number of samples (see NATO report [13] Annex B and [14]).

Signal Response (or â vs. a)

Ultrasonic inspection systems generally give a response whose amplitude depends on the characteristic parameter of the defect. In that case y is generally expressed in % of Full Screen Height (% FSH).

The amplitude information can be introduced in the POD curve estimation process. This methodology is called "signal response" or "â versus a". Compared to hit/miss analysis signal response analysis allows to estimate the parameters of the POD curve with a significantly lower amount of data.

The method assumes a linear model for the data corresponding to assumptions sometimes called "Berens hypothesis".

1. The measured data $y = \hat{a}$ (or $y = \log(\hat{a})$) are scattered around a linear trend versus a (or $\log(a)$)
2. the fluctuations are normally distributed and centered on the linear trend
3. the variance σ_ε are uniform on the entire range of defect characteristic parameter (homoscedasticity)

Fig. 8.4 The POD is the
fraction of the scatter density
that is above the detection
threshold

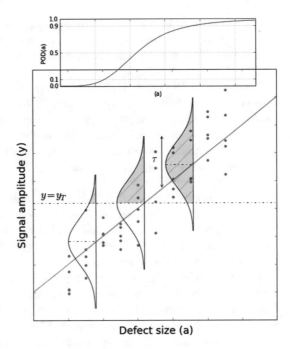

In practice, in the majority of the situations, at least one of the four combinations of â or log(â) versus a or log(a) correctly fits this linear model. The chosen combination is the one which exhibits the best fit (see ENIQ report 41, Sect. 3.3). The scattering of data around the general trend is due to the variability of the influential parameters captured by the trials and to the noise.

This linear model can be formalized as the following: $y = \beta_0 + \beta_1 g(a) + \varepsilon$ with $\varepsilon \sim N(0, \sigma_\varepsilon^2)$ and $g(a) = a$ or $g(a) = \log(a)$.

The POD curve can then be interpreted as the fraction of the density of the scattered data that are above the detection threshold y_T (see also Fig. 8.4).

In this framework the POD curve can be expressed by:

$$\text{POD}(a) = P(y > y_T) = 1 - \Phi_{norm}\left(\frac{y_T - (\beta_0 + \beta_1 g(a))}{\sigma_\varepsilon}\right) = \Phi_{norm}\left(\frac{g(a) - \mu}{\sigma}\right)$$

where $\Phi_{norm}(z)$ is the standard normal cumulative density function (probit link) and

$$\mu = \frac{y_T - \beta_0}{\beta_1}, \sigma = \frac{\sigma_\varepsilon}{\beta_1}$$

μ and σ are two parameters of the POD functional.

Like in the hit/miss analysis μ and σ can be interpreted as, respectively, the flaw parameter for which the POD is 50% and the steepness of the POD curve. The choice of one function in hit/miss analysis (logit) and another one in signal response analysis (probit) is historic, in order to ease the analytic computations.

The parameters μ and σ that best fit the data are determined using the maximum likelihood estimation methodology.

We refer to ENIQ report 41, Sect. 3.3.2 for the important discussions about the censoring of the data necessary to correctly take into account (very common) situations with saturation of the inspection system (the inspection system indicates 100% of FSH when the amplitude is above that value) or when the noise is not negligible (no amplitudes are recorded under the value of the noise).

We also refer to ENIQ report 41, Sect. 3.3.7 for the computation of the confidence bound of the POD curve.

When it appears that Berens hypothesis are violated, then another methodology must be applied. Different possibilities and tools exist and are described in the literature: application of hit/miss analysis,[1] Box-Cox transformation for data that violate homoscedasticity [30] or physics-based models [29] leading to a function g () that better fits the data for example.

It may be added that when the hypothesis funding the standard parametric estimation of the POD, are invalid, non-parametric methods can be investigated. Indeed the data required for carrying out reliable non-parametric estimation is prohibitive in the context of experimental POD. Simulation makes possible such analysis by providing sufficient set of data [32]. Nevertheless as emphasized in Part 4, Section "Step 4: Analysis of the Results and Computation of the POD Curve" the use of such non-standard methods, remains prospective and will not be discussed in this document.

Noise and Probability of False Alarms (PFA)

Other probabilistic criteria than POD curve such as Probability of False Alarm (PFA) or the Relative Operating Characteristics (ROC) curve are also important to characterize the performances of an NDT inspection system. PFA is the probability that the detection threshold is exceeded when no flaw is present (for instance, due to a strong background noise). An excessive PFA results in unacceptable rejection rates of the inspection process. Therefore a compromise must be found during the definition of the inspection process between PFA and POD (Fig. 8.5). A good knowledge of the noise properties is required to assess the PFA and fix an adequate threshold.

---

[1]Note that it is always possible to perform an hit/miss analysis from signal response data using:

$$\begin{cases} \hat{y} = 0 \; if \; y < y_T \\ \hat{y} = 1 \; if \; y > y_T. \end{cases}$$

Fig. 8.5 Influence of the detection threshold y_T on the PFA and the POD

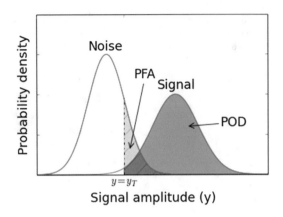

It is not the purpose of this document to give recommendations on how to determine PFA.

In the case of ultrasonic inspection the noise has two origins: electronic noise due to the acquisition chain (generally of low level if the system works correctly) and ultrasonic backscattering noise due to the material microstructure.

In some cases, ultrasonic backscattering noise level for the inspection process can be determined by simulation. However suitable models or input data to feed them are not always available. In that case, the noise level at the inspection frequency must be evaluated experimentally, this might be sufficient to find a detection threshold (generally higher than twice the peak noise) that gives an acceptable PFA.

In situations for which noise is important, it is crucial to correctly take into account its influence on the inspection results, especially through the censoring of the data in signal response analysis (see previous paragraph), and on the model used to fit the data (see next paragraph).

Min(POD) > 0 and/or Max(POD) < 1

In some situations, the classical two-parameter model described in the previous paragraphs (standard normal cumulative density function or logit) is not adapted. This can occur when the minimum POD does not go to zero (for example because of excessive background noise) or when the maximum POD does not go to one (resulting from random misses not related to target size). In this cases, the two-parameter model of the POD has not the same characteristics as the data since it enforces two asymptotes to POD = 0 and POD = 1; it must therefore not be used.

Three or four-parameter models have been developed to deal with these cases [30, 33]:

Fig. 8.6 Four-parameter
model with *lower* and *upper*
asymptote

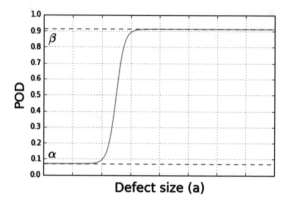

$$\text{POD}(a) = \alpha + (\beta - \alpha) \times \left[1 + \exp\left(-\left(\frac{g(a) - \mu}{\sigma}\right)\right)\right]^{-1} \quad \text{(logit)}$$

$$\text{POD}(a) = \alpha + (\beta - \alpha) \times \Phi_{norm}\left(\frac{g(a) - \mu}{\sigma}\right) \quad \text{(probit link)}$$

The parameter α is the lower asymptote and β is the upper asymptote of the POD
(Fig. 8.6). These two parameters can be included together into the model or only
one might be sufficient, leading to respectively four and three-parameter models.

Variation of Sensitivity in the Inspected Zone

When the inspected zone is large enough so that the variation of sensitivity of the
NDT system between different parts of the zone is significant, it is necessary to
extract from the analysis the influence of the position of the defect. The easiest way
to do that consists in dividing the inspected zones in different parts and to calculate
one POD curve for each part. An example of configuration for which this is necessary is presented in the appendix. Another way can be to calculate a
two-dimensional POD depending on both the size and the location of the defect
when the sensitivity mainly varies with one coordinate—depth—(multi-parametric
model).

Chapter 9
Best Practices Principles and Guidance for POD Curves Estimation Using Simulation

In this paragraph, the methodology and guidance to determine the POD curve by simulation is described.

The general methodology consists in using numerical models in order to reproduce the variability of the response of the NDT system induced by the variability of the influential parameters of the NDT inspection method (probe positioning, geometry, material,…). This is achieved by introducing uncertainties on the input parameters of the model to generate a large number of slightly different configurations, representative of real usages of the inspection system. The variability induced on the output of the model is statistically analyzed and used to calculate the POD curve. Such process is often called "propagation of uncertainties". This allows to accurately reproduce the variance observed on the response of the system even if a deterministic software is used for the simulations (Fig. 9.1).

The process for determining a POD curve using simulation of the inspection process should respect the following 6 steps:

1. To input in the simulation all the necessary information about the inspection (definition of one "nominal" configuration).
2. To determine and characterize the sources of variability (definition of aleatory parameters and assignation of statistical distributions).
3. To sample the statistical distributions of aleatory parameters (for example by Monte Carlo algorithm) and run the corresponding simulations.
4. From the set of simulated results to compute the POD curve.
5. To evaluate the reliability of the simulated POD curve.
6. To write a report with sufficient information to reproduce the results.

The second step is critical, since the POD curve obtained only reflects the influence of the selected sources of variability. The flowchart of Fig. 9.2 details the link between these different steps, which are described in the following paragraphs.

© Springer International Publishing AG 2018
B. Chapuis et al., *Best Practices for the Use of Simulation in POD Curves Estimation*, IIW Collection, https://doi.org/10.1007/978-3-319-62659-8_9

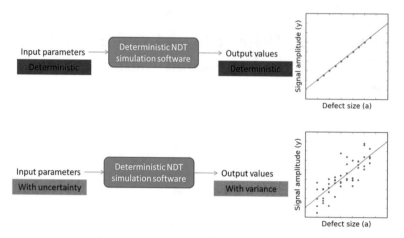

Fig. 9.1 Simulation of variance of the response using a deterministic software introducing uncertainty in the input parameters

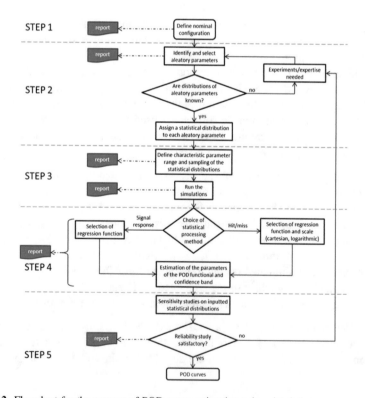

Fig. 9.2 Flowchart for the process of POD curve estimation using simulation

Step 1: Input of Nominal Values of Parameters

The first step consists in inputting in the software the set of parameters which will constitute the nominal inspection configuration from which will be defined statistical fluctuations.

For an ultrasonic inspection of welds the following parameters could be considered. However, this is neither an exhaustive nor a prescriptive list, and this must be adapted to the inspection and to the mathematical model which is used. A more exhaustive list can be found in the Appendix 1 of ENIQ RP1 [25].

- Specimen: The inputted parameters should enable to account for the propagation of the ultrasonic wave in the part. They concern:

  - The geometry
  - The properties of welded materials

    - Isotropy/anisotropy
    - Elastic properties (density, wave velocities …)
    - Description of the influence of ultrasound scattering by the microstructure: attenuation and noise

  - Environmental parameters such as: coupling medium, temperature (optional, if not known 20 °C is considered)

- Probe: The parameters should enable to compute the radiated beam and the sensitivity of the probe:

  - Crystal size and shape
  - Lens geometry and material
  - Frequency
  - Bandwidth
  - Wedge geometry and material
  - Focal laws if applicable

- Inspection system:

  - Excitation signal
  - Bandwidth, signal filters
  - Sampling frequency
  - Time gates
  - Gain
  - Detection thresholds

- Inspection parameters:

  - Position of the probe
  - Trajectory (range of displacement for manual UT)
  - Scanning speed (if applicable)

- Targeted defect:

  - Type (geometry, material)
  - Position and orientation

- Calibration reference:

  - Material and geometry of calibration block
  - Reflector type, position and geometry

The noise can have two origins:

- Interaction of ultrasonic beam with material microstructure of the component (structural noise),
- Electronic noise of the acquisition chain.

As described in Part 4, Section "Noise and Probability of False Alarms (PFA)" the noise has a strong influence on the Probability of False Alarm, and therefore fixes a minimal acceptable threshold for POD.

If the detection threshold is far above the tail of the noise distribution, the noise can be neglected. This choice must be included in the report (Part 4 section "Step 6: Report").

Otherwise, the noise must be included in the simulation. Two possible ways are the following.

- One is adding to the simulated signals a random noise following the statistical distribution (see Fig. 8.5) determined from experimental measurements. ENIQ report 41, Sect. 3.3.8 describes a procedure to obtain this distribution.
- When a physical model predicting the backscattered noise is available, the corresponding contribution can be directly included in the simulation. In such case, the electronic noise (generally of much lower amplitude) can be neglected.

In the first case, particular care must be taken to verify that the noise is constant within a component and from one component to another one. If it is not the case the noise should be considered as an aleatory parameter (see below).

Particular care should be taken to verify that the (graphical or not) user interface of the software allows to effectively account for the physical phenomenon that it is intended for. For example, the possibility to describe the anisotropy of the material (through the definition of different terms of the stiffness tensor) does not necessarily mean that the anisotropy is effectively modelled in the ultrasonic propagation.

Step 2: Description of Aleatory Uncertainties (Variability of Influential Parameters)

The second step consists in describing the variability of the influential parameters. This is done by assigning one statistical distribution to each parameter identified as aleatory (subject to variability).

- **Identification of aleatory parameters**

A pragmatic methodology to identify the aleatory parameters to include in the analysis has to be applied. Engineering understanding can be used to select or exclude from the statistical analysis some parameters. However, it is one of the advantages of simulation to make possible at low cost sensitivity studies. It is therefore recommended to perform as far as useful such sensitivity studies consisting in evaluating the influence of the variation of one parameter all other factors being kept constant. A parameter leading to a variation significantly lower than the other parameters on their expected (realistic) ranges of variation can be excluded of the statistical analysis.

In every case, for an ultrasonic inspection of a weld, the parameters listed in Part 4, Section "Step 1: Input of Nominal Values of Parameters" should be at least considered and the choice to keep every parameter or not in the analysis should be reported.

The simulated POD curve only reflects the influence of the selected aleatory uncertainties. The report (Part 4, Section "Step 6: Report") must therefore describe all the choices performed in this selection phase, in order to clearly define the perimeter of the conclusions that can be drawn from the POD curve.

- **Statistical distributions assigned to aleatory parameters**

The next step consists in describing the variability of the parameters identified as aleatory. This is done by assigning one statistical distribution to every aleatory parameter. In general the statistical distributions are defined by their functional form (normal, uniform, ...) and some associated parameters (mean value, variance, minimal and maximal value, ...). This information must be given in the report.

The range of variation of the statistical distribution must encompass all the values of the source of variability that could be encountered during the inspection due to the lack of control of this parameter.

To determine the most suitable statistical distributions assigned to one parameter the following sources can be used:

- Procedure (e.g. only the variation are given, not the distribution): however, on-field these values might not be respected. The purpose of a simulated POD study could be precisely to evaluate the tolerance of the procedure and its influence on the POD, but the value given in the procedure can be a first value to begin with.
- Engineering judgment (or questionnaire filled by on-field specialists, based on their experience)
- Direct measurement (for example measurement of the sample thickness variability to determine its distribution)

When information can be determined only on the variation range but not on the statistical distribution (and engineering judgment sometimes cannot provide an answer), then a uniform distribution on the entire variation range should be used to reflect the lack of information. This is a conservative choice if the support of this uniform distribution is large enough and encompasses all the variation range.

Step 3: Simulations

Once uncertainties have been characterized it is possible to define the set of numerical experiments (that is the set of simulations) that will be carried out and run computations according to a sampling algorithm such as Monte Carlo or derived method.

First of all, the values of the characteristic parameter are determined, as in an experimental study. The range of variation of the characteristic parameter must cover the expected domain of definition of the POD curve and especially the transition zone (corresponding to a rapid variation of the POD) which, in practice, is not known a priori. Some preliminary simulations can help to estimate the position of the transition zone for the given detection threshold.

The values of the characteristic parameter should be uniformly spaced on a Cartesian scale as recommended in the Sect. 4.5.2.2 "Target sizes and number of 'flawed' and 'unflawed' inspection sites" of the MIL-HDBK-1823A handbook. More detail is given in ENIQ report 47 about the repartition of defects in the hit/miss case [17]. The sampling of characteristic parameter values should be sufficiently fine to correctly describe the transition zone.

For each characteristic parameter value, a set of configurations is then randomly generated from the statistical distributions assigned to aleatory parameters. For each configuration, the generated values of the different aleatory parameters must be recorded in the report.

The simulations are then performed for all the configurations.

- **Recommended number of simulations**:

The simulations should be performed for a sufficiently large number of configurations, compatible with the computing resources available and confidence required. For example the MIL-HDBK-1823A suggests the following numbers for experimental studies:

- 60 for hit/miss analysis,
- 40 for signal response analysis.

This should give a POD curve with a reasonable confidence if the samples are correctly spread on the whole size range of defect size as described in ENIQ report 47 for hit/miss case (the same holds for the repartition of defect in the signal response case even if a complete study on this point as in ENIQ report 47 does not seem to have been carried out yet).

However, due to the efficiency of computer and simulation performances, the number of configurations used in practice will be in general much larger. This will reduce the problems of convergence of the maximum likelihood estimation or the strange behavior of the confidence curve that can be observed when only a limited number of configurations are used to compute the POD curve.

A large number of simulations should lead to a confidence band very close to the POD curve (sampling error minimized), which in practice gives $a_{90/95} \sim a_{90}$.

Again, as already mentioned in the document the confidence band only reflects the effect of the sampling. Indeed sampling is not the only factor governing the reliability of the estimated POD. One advantage of simulation is that it is possible to investigate the influence of epistemic uncertainty on the estimated POD. This can be done by performing sensitivity analysis on the POD curve as described in Part 4, Section "Step 5: Evaluation of the Reliability of a Simulated POD Curve". Such sensitivity analysis will provide information on the confidence which can be granted to the estimated POD.

Step 4: Analysis of the Results and Computation of the POD Curve

The simulated results can be used to calculate standard POD curves using the same methods used in experimental studies and described in ENIQ report 41 and MIL-HDBK-1823A.

- **Interpretation of the signal by an operator**:

If an automatic detection is not possible on the simulated signals or images for hit/miss analysis, a qualified operator should be used to analyze the results. This can be the case, for example, for TOFD results. In that case the results should be presented in the same conditions as on-field inspection: image size, contrast, brightness, contrast, color enhancement, ambient lighting, zooming possibilities,…

- **Statistical analysis and computation of the POD curve**

The hit/miss or signal response analysis should be performed as described in the ENIQ report 41. As explained in ENIQ report 41, the binomial approach (and the related "29 of 29 method") should not be used since the parametric modelling of the entire POD curve used for hit/miss or signal response analysis gives far better results with much more confidence. The confidence band should be determined using the methodology described in ENIQ report 41. For hit/miss analysis the confidence estimation using the loglikelihood ratio is recommended as justified in [13]. However, for large number of data (>200), the approximation given by Cheng [9, 11] is also convenient [13].

- **Choice between hit/miss and signal response analysis**

When the inspection system produces a quantitative response which respects the hypothesis described in Part 4, Section "Signal Response (or â vs. a)"

Since, simulation provides the amplitude of the ultrasonic signals (even when the real inspection system delivers a hit-miss binary data) it is always possible to perform a signal response analysis of the data to determine the POD curve. Provided that the Berens hypothesis (reported Part 4, Section "Signal Response (or â vs. a)") are verified, this analysis is more efficient (narrower confidence limit that accounts for statistical sampling error) for a given number of data: hit/miss analysis does not use the full information contained in the data in that case. Nevertheless simulation usually allows to obtain enough data to perform hit/miss analysis with a narrow confidence limit at low cost, a practical recommendation is to prefer hit/miss analysis. It is obviously especially the case when Berens hypothesis are not fulfilled.

- **Use of non-standard approaches**

It has been mentioned in Part 4, Section "Signal Response (or â vs. a)" that simulation makes possible to go beyond linear model and to apply non-parametric methods when the hypothesis funding the standard methodology are invalid. The use of such methods remains prospective and will not be discussed further in this document.

Step 5: Evaluation of the Reliability of a Simulated POD Curve

As described in Part 4, Section "POD Curves" the POD curves are given with a confidence band which is assumed to account for the uncertainty involved in the POD estimation process. Indeed in the classical approach (as described in MIL-HDBK-1823A or in ENIQ report) the confidence band only reflects the uncertainty due to the limited size of the dataset used for the POD evaluation. It has to be noticed that a lack of representativeness of the design of experiment is not accounted for by the confidence band as it is actually calculated. It has also to be recalled that the calculation of this confidence band is done assuming the statistical model (Berens hypothesis in the case of signal response analysis, functional form hypothesis for the POD curve) is valid.

When the POD is determined using simulation it is possible to make the confidence band as narrow as one wish by increasing the amount of simulated data. The width of the confidence band decreases when this number increases. The confidence band then converges to the POD curve, indicating that the lack of knowledge on the "true" value of POD due to a limited number of available data (sampling error) is no longer a problem.

However, this does not mean that the POD curve obtained by simulation is no more subject to uncertainty. There are different sources of errors or uncertainty which are listed in the following with associated recommendations.

- **Validity of the statistical model**

It is well known that the reliability of the POD estimated in the statistical framework recalled in Part 4, Chapter "Basics of Statistics for POD" strongly depends on the validity of the hypothesis done (Berens hypothesis in the case of signal response analysis). This remains true for simulation as long as the POD is estimated in the same statistical framework.

Therefore there are no recommendations specific to simulation for this point. It is needed to check that simulated data used for the POD estimation verifies statistical hypothesis exactly in the same way as for experimental data.

However, as mentioned above, by simulation it is possible to overcome these hypothesis since very large amount of data can be provided. Another methodology for the estimation of the POD can be employed (non-parametric approach [32]) in the aim of evaluating the influence of statistical hypothesis on the accuracy of the POD.

- **Representativeness of the design of experiments: Sensitivity to statistical distributions**

As mentioned before the lack of representativeness of the design of experiment is also a source of error on the POD curve estimation. In the case of POD obtained by simulation the design of experiment is done through the assignation of statistical distributions to the aleatory parameters. Indeed, the estimation of these statistical distributions is somehow delicate and in general results from an engineering judgment.

It has to be noticed that the question of representativeness exists also for purely experimental POD estimation and the design of experiments (the choice of the samples, of the defect distribution,…) implicitly assumes a certain knowledge on the real statistical distribution of the aleatory parameters. A lack of representativeness of the design of experiment may introduce a bias in the estimation of the POD curve, even if it is not always clearly mentioned. This bias is difficult to estimate through experimental campaigns. It is usually neglected compared to the uncertainty due to sampling error.

In the case of simulation studies it is possible and recommended to estimate the POD robustness linked to this point. This can be done by conducting a sensitivity study of the POD curve to the statistical distributions of the aleatory parameters. The recommended methodology is to identify the parameters of the statistical distributions inputted in the code which are not known with a good precision (interval bounds for uniform distributions or mean value, variance, for normal distributions…) and to launch various POD estimation assuming extreme values for these parameters. If this sensitivity study exhibits unacceptable dispersion on the POD values it is necessary to initiate actions allowing to obtain more information about the unknown parameters.

A more systematic methodology has been proposed in Ref. [16], to quantify the uncertainty induced on the POD values by the lack of knowledge on statistical distributions. In this approach the POD is itself considered as a random variable due to this uncertainty. The methodology consists in:

- Defining a probability distribution for the parameters of the statistical distributions inputted in the code which are not known with accuracy. These probability distributions reflect the level of knowledge on the various parameters.
- Sampling these probability distributions (by Monte Carlo method). To each sample corresponds one statistical distribution for every aleatory parameter.
- Computing the POD corresponding to every sample.

The result is a collection of POD curves which can be statistically analyzed. In particular it is possible to determine the mean of the POD and tolerance intervals on the POD estimation. The information given by such tolerance interval at $\alpha\%$ on POD(a) is the following: Considering the level of knowledge on the statistical distributions, the probability that the «true» POD (true if the other hypothesis of the calculation were perfectly verified) is higher than the lowest bound of the interval is $\alpha\%$.

This methodology is recommended when the statistical distributions are not known with precision and when it is practically tractable. However, this is not always the case since it requires huge amount of computations. Research is in progress in the community to reduce the cost of such systematic methodology.

- **Validity of the physical model**

A model never gives "perfect" predictions. Its validity and the representativeness of its input are always limited. The accuracy of the model has to be checked through experimental validations as discussed in Part 3, Chapter "Validation of the Software Tool". It has been recalled that it exists two possible causes of errors and therefore uncertainty on simulated data:

- Uncertainty due to the validity of the model and its implementation,
- Uncertainty due to a limited knowledge on the inputs of the model (independently of the variability on uncontrolled parameters considered in the POD estimation).

However, if the experimental validation of the simulation is done on the materials perfectly representative of the inspected components and using the same inspection devices as for real inspections then there is no need to distinguish between these two possible causes of errors. The uncertainty on the input of the model is included in the accuracy of the model measured by the experimental validation. If it is not the case, the uncertainty on the input of the simulation has to be considered.

The influence of the model errors on the POD strongly depends on how the error varies with inspection parameters and in particular with the size of the defect. This influence will also be very different if the error includes some systematic bias or if it can be considered as randomly distributed around 0. Depending on the situation, the model error can affect the value of the estimated POD and/or the confidence on this value.

The development of a methodology to include the model error in the analysis remains a subject of research. The recommendations are given here with the present stage of understanding on this question and in an aim of conservatism:

Let us consider that the uncertainty on the predictions of the model is expressed by an interval (e.g. $\pm$ 3 dB) reporting the maximum of the discrepancy between experiment and simulation which has been measured during the validation campaign.

In ultrasonic applications we can consider that when interval of errors is close to $\pm$ 2 dB (comparable to the experimental uncertainty), simulation can be used with a high level of confidence.

If the interval of errors is higher than $\pm$ 3 dB simulation has to be used with more care and it is recommended to take into account the model error in a conservative way. This can be done by considering that an interval of error wider than $\pm$ 3 dB (let say $\pm$ (3 + δ) dB) indicates the possibility of a systematic bias of δ dB. Such bias corresponds to the "worst case". A conservative way is to estimate the POD after having substracted to all the simulated data the value of δ dB or, which is equivalent, to estimate the POD with a threshold of detection augmented by δ dB.

Step 6: Report

A report presenting the results should mention all the parameters that could allow reproducing the simulations and the statistical analysis:

- Identification of the used software (names and versions)
- All the data of Part 4, Section "Step 1: Input of Nominal Values of Parameters"
- The list of the aleatory parameters and the assigned statistical distributions
- Calibration value (sensitivity)
- Detection and saturation thresholds
- Type of analysis (hit/miss or signal response)
- Type of regression in hit/miss analysis (cumulative log-normal, logit, …)
- Type of linearization in signal response analysis (log-log, lin-lin, …)
- Results of the simulations.

The limitations of the conclusions that can be deduced from the simulated POD curve should also be described:

- List of (potential) influent factors that have not been included in the simulation
- Justification for factors for which influence has been considered negligible
- Results of reliability study on the simulated POD curve (as per Part 4, Section "Step 5: Evaluation of the Reliability of a Simulated POD Curve").

Chapter 10
Best Practice Principles and Guidance for the Exploitation of Simulated POD Data in Relation with Experimental Data

Complement an Experimental Data Set for Estimation of a POD Curve

The methodology and guidance presented in the previous part allows to estimate a POD curve fully based on simulations. However, in some cases, it might be interesting to benefit from the advantages of both experiments and simulations to determine a POD curve with larger amount of data. Indeed, the variability induced by some aleatory parameters, difficult to model, can be captured experimentally. In contrast, adding simulated data to an experimental POD curve can be used to complement a reduced number of available experimental data.

In practice:

- It is possible to add simulated data in order to estimate a POD while the number of available experimental data is too reduced to verify the criteria of the MILHDBK 1823. The reliability of the POD curve complemented by simulation has to be evaluated in the same way as described above in Part 4 section "Step 5: Evaluation of the Reliability of a Simulated POD Curve".
- However, it is important to be aware that the reduction of the confidence band consecutive to the adding of simulated data is somehow artificial and is not a measure of the reliability of the estimated POD.

Two possibilities are proposed in literature to mix data coming from different nature to obtain a POD curve. The transfer function approach has been extensively studied to transfer POD measured for one specific application or set of conditions to another related application using experimental data and models [21].

More recently, the Bayesian approach proposes a general framework to update the state of knowledge of the POD curve as long as more data (experimental or simulated) are added [19, 20, 29]. This approach offers the benefit of a large versatility and an easy application within a rigorous mathematical framework.

© Springer International Publishing AG 2018
B. Chapuis et al., *Best Practices for the Use of Simulation in POD Curves Estimation*, IIW Collection, https://doi.org/10.1007/978-3-319-62659-8_10

However, these approaches remain matter of research and no examples for UT inspection of welds seem to have been presented yet, either using the transfer function approach or the Bayesian approach.

Comparison Between Experimental POD Curves and POD Curves Estimated by Simulation

Comparing POD curves obtained for the same configuration fully by experiments and fully using simulation requires care. It is necessary to avoid the risk of comparing data of different natures. This risk comes from the different analysis and statistical methodologies that are used to calculate the POD curve as well as the general lack of information on what and how it has been done.

A guideline written by the Australian DSTO identifies the most relevant questions that must be addressed when comparing two POD curves, mainly with the objectives of comparing (experimental) data published in literature [6].

For the specific case of comparing POD curves obtained experimentally and by simulation, the following aspects must also be taken into account:

- Since in practice the performance of the NDT system is quantified by the $a_{90/95}$ value, the same statistical method should be used to estimate the confidence band. In particular, to compare the confidence band (or $a_{90/95}$), the same number of points must be considered, even if simulation generally allows to use much larger dataset.
- It is recommended to use the same statistical software to process both experimental and simulated data to calculate the respective POD curves. This requires having access to the raw data, which is, unfortunately, not always the case.
- The same influential factors must have been controlled both experimentally and in the simulations (in particular, the human factors must have been proven to be negligible in the experimental study if they have not been included into the simulation).

Differences between experimental POD curve and POD curve determined by simulation does not necessarily indicate that the simulations are not correct. The two approaches have advantages and drawbacks and the possible sources of inaccuracy are not the same. Further analysis must be carried out to understand the sources of discrepancies.

Simulations and experiments have relative advantages and drawbacks.

The main advantages of experiments are the following:

- If the testblocks are perfectly representative, the experimental POD may account for unknown or not perfectly known characteristics of the part, the defect, etc.
- If the dataset is sufficient it may be assumed that the total inspection variability will be correctly taken into account in the experimental POD calculation.
- By definition there is no need to use models which are always subject to errors.

The main advantages of simulation are the following:

- The simulation offers the possibility to explore more extensively the perimeter of variations of the influential parameters.
- The simulations allow to reduce as far as one wishes the uncertainty on the estimation induced by limited sample size.
- By simulation it is possible to estimate the sensitivity of the POD estimation to the design of experiments (by modifying the statistical distributions of influent parameters).
- Simulation gives means to propose alternatives when Berens hypothesis are not verified.

Optimization of the Design of Experiments in a POD Study

As emphasized in Sect. 4.5.2.2 of the MIL-HDBK-1823A, it is crucial to distribute correctly the size of flaw in the experimental specimens to obtain more precise estimate of the statistical parameters.

> Given that $a_{90/95}$ has become a de facto design criterion it may be more important to estimate the 90th percentile more precisely than lower parts of the curve. This can be accomplished by placing more targets in the region of the a_{90} value but with a range of sizes so the entire curve can still be estimated.

A simulation study should allow determining the position of the transition zone of the POD curve before manufacturing the flawed specimens, and therefore target the distribution of flaw sizes (reduce number of tests in ranges where POD = 0 or 1). This consequently results in fewer testblocks necessary to determine the POD curve.

ENIQ report 47 provides interesting guidance on how to space the defects size in the hit/miss case when the transition zone has been determined [17].

Quantification of the Impact of an Additional Parameter

If, after the realization of a first POD campaign, the influence of an additional parameter must be evaluated, the simulation can be used in three different types of situation:

- The parameter has been neglected, it is required to justify that its influence is really negligible on the POD curve;
- The parameter can be controlled (and is probably not negligible) but it has not initially been considered, it is required to evaluate its influence;

- An aleatory parameter is in reality better controlled than initially planned (and for which the qualification program might have already been performed), it is wanted to claim better performances of the system.

Some complementary experiments could act as a partial justification in these different situations, with the risk to limit the experiments to the ones that effectively show the effect that it is wanted to prove (negligible influence for example). Rigorously, a new experimental program on a full test matrix should be performed, which is rarely done in practice for time and budget constraints. A more complete justification can be given by a full simulation study which easily allows to add some aleatory parameters in a full test matrix analysis.

If the first campaign has already been performed using simulation and the configurations have been stored, the quantification of the impact of the additional parameter can be very fast and easy to setup. On the contrary, if the first campaign has been performed experimentally the results cannot be directly reused (neither if it is wanted to perform a new experimental campaign nor for a simulation campaign). However, simulation can be validated by comparison with the first set of experimental results. It is then easy to compute the POD curve with the additional parameter without performing a full experimental study.

Example: The weld temperature influence might have been neglected during the qualification process. However, on field it can appear that for productivity reasons the inspection is often performed just after welding, when the weld is still hot. Simulation can allow to evaluate this factor and the POD curve that takes into account its influence without performing a new complete experimental program that includes this parameter.

Quantification of the Influence of a Change in the Inspection Procedure

The simulation can be a useful technical justification in the case of a change in the inspection procedure. Two things must be verified:

- The simulation must correctly reproduce the experimental results (and the POD estimation) of the initial case (Part 4 section "Comparison Between Experimental POD Curves and POD Curves Estimated by Simulation").
- If new aleatory parameters must be evaluated the validity of the model must be checked on the entire range of variation of the new aleatory parameters (Part 3 chapter "Validation of the Software Tools").

In that case, a new test matrix must be generated to reflect the changes of the procedure (new aleatory parameters, adjustments of the statistical distributions of the initial ones) and the simulation run. The resulting POD curve can be compare to the initial one to quantify the influence of the change in the inspection procedure.

Example: if the acceptance criteria become more stringent (i.e. it is required to detect smaller defects), a possibility could consists in using a probe at higher frequency and test the possibility to detect smaller defects (the threshold must be adapted to take into account a probable higher noise in that case). Simulation could allow to avoid to perform a completely new experimental campaign. However, it is important to verify that the simulation models are valid at this frequency.

Modify the Procedure to Bring Back a POD Curve to an Acceptable Value

Simulation easily allows to determine the most influential parameters that degrade the performances of the inspection technique. Indeed, a sensitivity study can be performed to identify the parameters that mostly contribute to a shift of the POD curve. Once these parameters have been identified, it is possible to directly determine to what extent it is necessary to control them to guarantee an acceptable value of the POD and to update the procedure accordingly.

Example: In an inspection procedure, the probe must generally be positioned at a nominal point on the specimen with a given tolerance. If it appears that this parameter has a bigger impact on the POD than other factors, it might be interesting to be more stringent on its tolerance to improve the POD. Simulation easily allows to quantify it, and to predict the expected performances of the inspection system of the accordingly modified procedure.

Conclusion

The methodology and guidance presented in this document provides a framework for the determination of POD curves using simulation of the inspection technique. Although the current document focuses on the simulation of the UT weld inspection technique, many of the same principles could be applied to other techniques and situations. The statistical framework to compute the POD curves is exactly the same as those generally used in qualification programs and described in the reference documents MIL-HDBK-1823A and ENIQ Report 41. The requirements in terms of validation of the software and competences of the user have been specified and different possibilities to use the results of such simulations have been listed.

The research on this subject is now mature enough to consider systematic applications in industrial qualification programs. Due to the versatility of the simulation tools and the several possibilities offered (time and cost savings, better reliability, possibility to explore configurations not possible experimentally, systematic tool to optimize procedure to guarantee a POD), it should provide an high value in a POD estimation process.

The full potential of the use of simulation in qualification programs will be provided by the approach that consists in mixing the results of a large number of simulations with a limited number of experimental data to obtain a POD curve with a better reliability at a reduced cost. Successful applications of this approach in the field of UT weld inspection still need to be published. However, the research in this field is very active and formal frameworks and methodologies to perform it rigorously exist in the literature and should easily be applied to UT weld inspection.

© Springer International Publishing AG 2018
B. Chapuis et al., *Best Practices for the Use of Simulation in POD Curves Estimation*, IIW Collection, https://doi.org/10.1007/978-3-319-62659-8

Appendix
Specificities of AUT Systems Using Zonal Discrimination Method

The inspection of pipeline girth welds is commonly performed using Automated Ultrasonic (AUT) systems composed of several mono-element transducers on both sides of the weld or two facing phased array probes with several focal laws. These systems allow, with a single scan along the circumference of the pipe, a complete examination of the weld using the zonal discrimination approach.

The zonal discrimination approach is described in the ASTM E1961 standard practice [9]. It consists in dividing the thickness of the pipe wall in different zones of 1–3 mm height corresponding approximately to stringer bead height (Fig. A.1). Each zone is inspected using a dedicated ultrasonic beam (one mono-element probe ore one focal law) directed at a fixed position usually in pulse-echo or tandem (pitch-catch) configuration. These dedicated ultrasonic beams are usually called "channels". The ASTM E1961 standard practice requires a minimal (and maximal) overlying between two channels to avoid dead zones of the inspection system. Supplementary TOFD channels can also be added to increase the performances of the system.

The amplitude of the echo from a defect is compared to that of calibrated reflectors (flat bottom holes or notches) associated to each zones. This allows to locate the depth of the defect and to give an approximation of its height (in terms of number of channels that detected the defect).

The demonstration of performances of such systems is generally performed according DNV guidelines which require the estimation of POD curves (and sizing accuracy) [1, 2].

Since the system is composed of several channels, a special care must be taken to determine the reliability of the inspection system. In particular, the analysis must be performed zone per zone in order to avoid situations for which bad performances in one zone of the weld are counterbalanced by very good detection in other zones. For each zone, the number of configurations simulated must respect the conditions given in Part 4, Section "Step 3: Simulations".

© Springer International Publishing AG 2018
B. Chapuis et al., *Best Practices for the Use of Simulation in POD Curves Estimation*, IIW Collection, https://doi.org/10.1007/978-3-319-62659-8

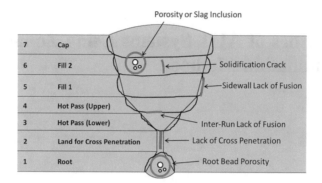

Fig. A.1 Typical zones and kind of researched defects

Different channels can detect the same defect (due to the overlying of ultrasonic beams between two adjacent channels or thanks to the use of a supplementary TOFD channel), the POD curve for one zone can therefore be obtained by mixing the detection results for the different channels that might detect defect in that zone. This is only possible using a hit/miss analysis and considering, for each zone, a subsystem constituted from the different probes that might possibly detect defect in that zone. If one probe sees the defect, it means that the whole subsystem sees the defect, which leads to a better POD for that zone compared to that obtained considering only the nominal channel.

It should be noted that mixing information coming from different probes is not directly possible for signal response analysis (however, it is always possible to perform a hit/miss analysis from signal response data, see Part 4, Section "Signal Response (or â vs. a)").

The performances that are claimed for the whole inspection process should be based on the worst case obtained considering all the different zones of the weld or detailed zone by zone.

References

1. DNV-RP-F118. (2010). Pipe girth weld AUT system qualification and project specific procedure validation, Recommended practice, Det Norske Veritas.
2. Forli, O. (1999). Guidelines for NDE reliability determination and description. *DNV NordTest Technical Report 394.*
3. MIL-HDBK-1823A. (2009). Nondestructive evaluation system reliability assessment, *US Department of Defense.*
4. Georgiou, G. (2006). Probability of Detection (PoD) curves—Derivation, applications and limitations, *Health and Safety Executive.*
5. Dominguez, N., & Jenson, F. (2010). Simulation-supported POD. In: *AIP Conference Proceedings on Review of Quantitative Nondestructive Evaluation.*
6. Harding, C., & Hugo, G. (2011). Guidelines for interpretation of published data on probability of detection for nondestructive testing, *DSTO.*
7. Calmon, P. (2013). *Recommendations for the use and validation of non-destructive testing simulation*, International Institute of Welding.
8. Gandossi, L., & Annis, C. (2010). Probability of detection curves: Statistical best-practices, ENIQ report 41.
9. ASTM-E1961-11. (2011). Standard practice for mechanized ultrasonic testing of girth welds using zonal discrimination with focused search units.
10. Cheng, R., & Iles, T. (1988). One-sided confidence bands for cumulative distribution functions. *Technometrics, 30,* 155–159.
11. Cheng, R., & Iles, T. (1983). Confidence bands for cumulative distribution functions of continuous random variables. *Technometrics, 25,* 77–86.
12. Berens, A. (1989). NDE reliability data analysis, *Nondestructive Evaluation and Quality Control*, 17, 689–701.
13. RTO-TR-ATV-051. (2005). *The use of in-service inspection data in the performance measurement of non-destructive inspections*, NATO.
14. Harding, C., & Hugo, G. (2003). Statistical analysis of probability of detection hit/miss data for small data sets. In *AIP Conference Proceedings on Review of Quantitative Nondestructive Evaluation*, 657.
15. Annis, C. (2014). Influence of sample characteristics on probability of detection curves. In *AIP Conference Proceedings on Review of Quantitative Nondestructive Evaluation.*
16. Dominguez, N., Reboud, C., Dubois, A., & Jenson, F. (2013). A new approach of confidence in POD determination using simulation. In *Review of Quantitative Nondestructive Evaluation (Vol 32B)*, AIP Conference Proceedings (Vol 1511, 1749–1756).
17. Annis, C., & Gandossi, L. (2012). Influence of sample size and other factors on hit/miss probability of detection curves. *ENIQ Report, 47.*
18. Thomson, B. (2007). A unified approach to the model-assisted determination of probability of detection. In *AIP Conference Proceedings Review of progress in Quantitative Nondestructive Evaluation.*

© Springer International Publishing AG 2018
B. Chapuis et al., *Best Practices for the Use of Simulation in POD Curves Estimation*, IIW Collection, https://doi.org/10.1007/978-3-319-62659-8

19. Jenson, F., Dominguez, N.; Willaume, P. & Yalamas, T., A Bayesian approach for the determination of POD curves from empirical data merged with simulation results. In *AIP Conference Proceedings on Review of progress in Quantitative Nondestructive Evaluation.*

20. Kanzler, D., Müller, C., & Pitkänen, J. (2013). Evaluation of radiographic testing performance with an advanced POD approach. In *5th European-American Workshop on Reliability of NDE.*

21. Harding, C., Hugo, G. R., & Bowles, S. J. (2009). Application of model assisted POD using a transfer function approach. In *AIP Conference Proceedings on Review of progress in Quantitative Nondestructive Evaluation.*

22. Bertovic, M., Fahlbruch, B., Müller, C., & Pitkanen, J., Ronneteg, U., Gaal, M., Kanzler, D., Ewert, U., Schombach, D. (2012). Human factors approach to the acquisition and evaluation of NDT data. In: *Proceedings of the 18th World Conference on Nondestructive Testing,* Durban, South Africa.

23. http://www.hse.gov.uk/humanfactors/.

24. Müller, C., Holstein, R., & Bertovic, M. (2013). *Conclusions of the 5th European-American Workshop on Reliability of NDE.* http://www.nde-reliability.de/portals/nde-reliability2013/BB/conclusion.pdf.

25. ENIQ. (2005). Recommended practice 1 influential/essential parameters (*ENIQ* Report 24 EUR 21751 EN).

26. Li, M., Spencer, F. W., & Meeker, W. Q. (2012). Distinguishing between uncertainty and variability in nondestructive evaluation. In *AIP Conference Proceedings on Review of progress in Quantitative Nondestructive Evaluation.*

27. Der Kiureghian, A., & Ditlevsen, O. (2009). Aleatory or epistemic? Does it matter? *Structural Safety, 31,* 105–112.

28. http://www.hse.gov.uk/research/rrpdf/rr617.pdf.

29. Aldrin, J. C., Knopp, J., & Sabbagh, H. A. (2013). Bayesian Methods in Probability of Detection Estimation and Model-assisted Probability of Detection (MAPOD) Evaluation. *Review of Progress in QNDE, 32,* 1733–1740 (AIP).

30. Knopp, J. S., Ciarallo, F., & Grandhi, R. V. (2015). Developments in probability of detection modeling and simulation studies. *Materials Evaluation, 73,* 55–61.

31. Annis, C., Aldrin, J. C., & Sabbagh, H. A. (2015). What is missing in nondestructive testing capability evaluation? *Materials Evaluation, 73,* 44–54.

32. Spencer, F. W. (2011). Nonparametric Pod estimation for hit/miss data: A goodness of fit comparison for parametric models. In *AIP Conference Proceedings on Review of Quantitative Nondestructive Evaluation.*

33. Annis, C., Aldrin, J. C., & Sabbagh, H. A. (2015). Profile likelihood: What to do when maximum probability of detection never gets to one. *Materials Evaluation, 73,* 96–99.